INSIDE MODERN GENETICS

INSIDE THE BASICS OF GENETICS

HOWARD PHILLIPS

Published in 2022 by The Rosen Publishing Group, Inc.
29 East 21st Street, New York, NY 10010

First Edition

Portions of this work were originally authored by Carol Hand and published as *Introduction to Genetics*. All new material in this edition was authored by Howard Phillips.

Designer: Rachel Rising
Editor: Greg Roza

Cataloging-in-Publication Data

Names: Phillips, Howard, 1971-.
Title: Inside the basics of genetics / Howard Phillips.
Description: New York : Rosen Publishing, 2022. | Series: Inside modern genetics | Includes glossary and index.
Identifiers: ISBN 9781499470437 (pbk.) | ISBN 9781499470444 (library bound) | ISBN 9781499470451 (ebook)
Subjects: LCSH: Genetics--Juvenile literature. | Genes--Juvenile literature. | DNA--Juvenile literature.
Classification: LCC QH437.5 P54 2022 | DDC 576.5--dc23

Manufactured in the United States of America

CPSIA Compliance Information: Batch #CWRYA22. For further information contact Rosen Publishing, New York, New York at 1-800-237-9932.

CONTENTS

INTRODUCTION

Our world is filled with an incredible variety of plants, animals, and microscopic creatures. They have a wide range of physical traits. Even two creatures of the same species can look very different. Also, why do some twins look alike and others look different? How can two brown-eyed parents have a blue-eyed baby? People have been asking questions like these for thousands of years. These types of questions—and the search for their answers—eventually led to the development of the science of genetics.

Genetics is the study of heredity. This is the passing of traits from one generation to the next. Geneticists figure out how traits are inherited and determine how likely a trait is to show up in the next generation. The field of genetics covers anything to do with genes, the tiny units inside cells that make heredity possible. Genes are parts of a molecule of DNA, so genetics is also about DNA.

This is a visualization of a genomic analysis. This is just one way genomic data can be displayed and studied.

Why does genetics matter? Genetics has become essential in our understanding of health and disease. Scientists are discovering which diseases are inherited and how. We are discovering how likely it is that a child will inherit a condition from a parent or grandparent. Can scientists "fix" a gene so that future generations won't inherit the disease? Our supermarkets are stocked with a variety of healthy foods because of genetics work. Scientists are now using genetics to save endangered species and clean up polluted areas. They're even using genetics to solve crimes. More applications for the study of genetics are discovered all the time.

Genes are passed down from one generation to the next. Offspring often look a lot like their parents and grandparents.

CHAPTER 1

CHARLES DARWIN AND GREGOR MENDEL

About 10,000 years ago, groups of nomadic wanderers gave up the hunting-gathering lifestyle and settled down in specific areas. In time, they began to grow and store food. These early farmers domesticated crops and animals for food. They brought animals in from the wild and gave them food and shelter. They planted fields of grain and gardens of vegetables. Early farmers tried to improve their plants and animals by selective breeding. That means they selected the parents of each new generation. They chose grains with the fullest heads. They chose cattle and goats that produced the most milk. They believed the parents would pass these chosen characteristics on to new generations, making them more valuable.

For centuries, early farmers did not know exactly how parents' traits were passed on to their offspring. Farmers just selected good parents and hoped for the best. The babies often had the desired traits—but not always. People did not understand why traits sometimes showed up in the next generation and other times didn't.

Wheat was one of the first crops that people planted and cared for. Other early agriculture crops include barley, peas, lentils, and flax.

BEFORE GREGOR MENDEL

Around 400 BCE, the Greek philosopher Hippocrates theorized that seeds or particles produced in the bodies of both parents became blended in offspring. He suggested that this explained why a child showed characteristics of both parents. Another Greek philosopher, Aristotle, did not agree. He said it could not be true because some children don't resemble either of their parents. In the end, however, Aristotle's theory said much the same thing. He thought heredity occurred when fluids from both parents' bodies were mixed.

Hippocrates and Aristotle simply thought about how heredity occurred. They didn't make controlled observations, do experiments, or collect data. They were basically guessing, and they had no true evidence to back up their ideas. Nevertheless, people accepted the idea of "blending inheritance" for more than 2,000 years.

Another rather strange theory of heredity, known as preformation, arose in the 1700s and 1800s. This is the belief that future generations are already fully made within sex cells. They thought smaller and smaller organisms fit inside each existing organism, like tiny nested dolls. Using early microscopes, some people were convinced they could actually see tiny complete "men" inside human sperm cells.

THE DAWN OF GENETICS

Our understanding of modern genetics developed quickly in the 19th century when two scientists separately collected and analyzed large amounts of data. They developed theories to explain what they saw rather than guessing. These two scientists never worked together. They never even met. They worked in different countries and in different fields of science. After they died, however, other scientists connected their data and ideas. Together, their ideas formed the basis of modern genetics. One of these men, Charles Darwin, developed the theory of evolution. The other, Gregor Mendel, is now known as the "father of genetics."

Darwin was a naturalist and geologist. He was hired as the naturalist aboard the British survey ship HMS *Beagle* from 1831 to 1836. During this voyage around South America, Darwin collected numerous samples—plants, animals, and fossils—at every stop. He made detailed notes and observations about their structures and habitats. He noticed that organisms varied by habitat. South American plants and animals

Charles Darwin is shown here in 1880 as an old man, long after he had made his famous fact-finding trip around the globe.

looked different from the same types he was familiar with in Europe. At that time, most people believed that the living things on Earth were the same as they had always been. Darwin began to question this belief. His observations led him to think that life constantly changed, or evolved. His examinations of living things showed that they were adapted to their environments. His observations of fossils suggested that new species could evolve over very long periods of time.

Darwin continued to observe nature for many years after he returned to England, studying a variety of creatures including orchids, barnacles, and earthworms. He figured out patterns in his data and tried to explain them. Finally, in 1859, he published *On the Origin of Species*, which described his theory of evolution. Although another naturalist named Alfred Russel Wallace independently developed the same theory, it bears Darwin's name because Darwin published his observations and theory before Wallace did. Both men backed up this theory with many years' worth of evidence. The theory bears Darwin's name because he did far more research than Wallace.

Darwin became fascinated by the star orchid, which has a long green appendage that holds nectar. Darwin theorized that there was an animal capable of reaching that nectar, perhaps a moth with a very long proboscis. Twenty-one years after Darwin's death, scientists discovered that moth: the sphinx moth, sometimes called the hummingbird moth.

THE THEORY OF NATURAL SELECTION

Darwin's theory of evolution deals with changes within a population, which is a group of organisms of the same species living in the same place. Evolution describes changes to populations over long periods of time. The changes are very tiny, but they occur over many generations. Darwin's theory of natural selection proposes a process by which these changes occur.

A theory is a well-tested, evidence-based explanation of a natural phenomenon. It describes a process by which the phenomenon came to be. Darwin's theory of natural selection describes how populations adapt as their environment changes. Natural selection relies on two factors. One factor is the presence of inherited variations, or differences, which all organisms have. These variations are obvious in family members. They look similar, and you can tell they're related, but they're not identical. The other factor is stress on a population caused by changes in the environment. Environmental stress can include the lack of food or the presence of competing species or predators. A change in climate can also cause environmental stress.

According to Darwin's theory, all species undergo a "struggle for existence." For example, let's say some plants in a population are able to withstand hot, dry conditions better than others. In hot, dry years, these plants still reproduce and leave many new plants to form the next generation. Plants that can't withstand these conditions as well will die more quickly in such weather and produce fewer offspring. Therefore, the next generation has more plants that can survive these hot, dry conditions. If the weather stays hot and dry for many years, the "hot, dry" trait builds up. Plants that don't have it decrease in number. If most years are mild, both traits will be present in more equal numbers.

Darwin's observations described changes in traits through generations and convinced him that natural selection occurred in populations. However, he could not explain how traits were transferred between individual organisms. Someone had to answer this question before other scientists would accept natural selection as a valid theory. Darwin tried to explain it by pangenesis, or "whole birth." He suggested body cells shed tiny pieces called gemmules, which collected in reproductive organs and helped determine the offspring's characteristics. But Darwin had no evidence for pangenesis, and the mechanics of natural selection remained a puzzle.

Giraffes did not always have the extra long necks they have today. Scientists think they evolved longer necks to reach leaves higher in trees, thus increasing their chance of survival.

THE FATHER OF GENETICS

One missing piece of Darwin's puzzle was published in 1866, only seven years after his own work. This paper, "Experiments in Plant Hybridization," was written by an Augustinian friar named Gregor Mendel. Mendel's work proposed a method for passing traits from parent to child and gave evidence for it.

Johann Mendel was born on July 22, 1822, in Heinzendorf, Austria, a village that is now part of the Czech Republic. In 1843, Mendel entered the Abbey of St. Thomas in Brno and took the name "Gregor." The abbey was the perfect place to study natural science. It had scientific instruments, a large botanical collection, and a library with several thousand books. It also had extensive gardens. The friars, who were teachers, university professors, and researchers, encouraged Mendel's

Gregor Mendel's experiments with pea plants and his findings were not respected during his life.

interest in science. In 1851, the abbot sent Mendel to the University of Vienna. There, he studied physics and botany. Physics taught him to use experiments and mathematics. Botany stimulated his interest in plant variation and inheritance.

Mendel taught physics when he returned to the monastery, but he spent most of his time in the monastery gardens. He conducted studies of inheritance in garden pea plants from 1856 to 1865. Mendel's one published paper was ignored during his lifetime. But it was rediscovered in 1900, 16 years after his death.

THE LEGACY OF DARWIN AND MENDEL

The work of Charles Darwin and Gregor Mendel changed science forever. Darwin proved that populations and species often adapt to better live in their environments. Mendel proved that these adaptations are passed from generation to generation in a precise, predictable way.

Darwin was a master of observation. He closely studied nature to answer specific questions. He also meticulously recorded dates, times, locations, and weather conditions—any factor that might affect a population. Darwin knew it was important to continue observations for long periods of time. He tried to fit all his data into meaningful patterns, and he developed theories to fit them.

Mendel was an excellent observer too. But Mendel did more. He conducted experiments and analyzed them mathematically. He applied physics techniques to his biological studies. Mendel's precise experiments showed patterns in his data that led to the laws of genetics.

PASS THE PEAS!

Mendel chose garden pea plants as a test species to show inheritance patterns. Garden peas had many traits that were present in two—and only two—forms. These "either-or" traits were easy to distinguish by looking at the peas. Seeds were either yellow or green. Pea pods were either yellow or green. Some seeds were round and smooth while others were wrinkled. Plant stems were either long or short. Mendel studied seven pairs of these simple traits.

Peas can self-fertilize because the flowers contain both male and female parts. New plants form when a plant's pollen fertilizes its own egg cells. Mendel studied only pea plants that "bred true." That is, the offspring of self-fertilized plants always looked exactly like their parents. This meant Mendel could be sure of the ancestry of each plant.

Another reason Mendel chose peas was because he could pollinate them by hand. First, he removed the stamens from a flower before it produced pollen in order to prevent the plant from pollinating itself. Then he used a small paintbrush to dust mature pollen from another plant onto the flower. This cross-pollination produced hybrid plants.

Over the course of eight years, Mendel cultivated and tested between 28,000 and 30,000 pea plants. He ensured all pea

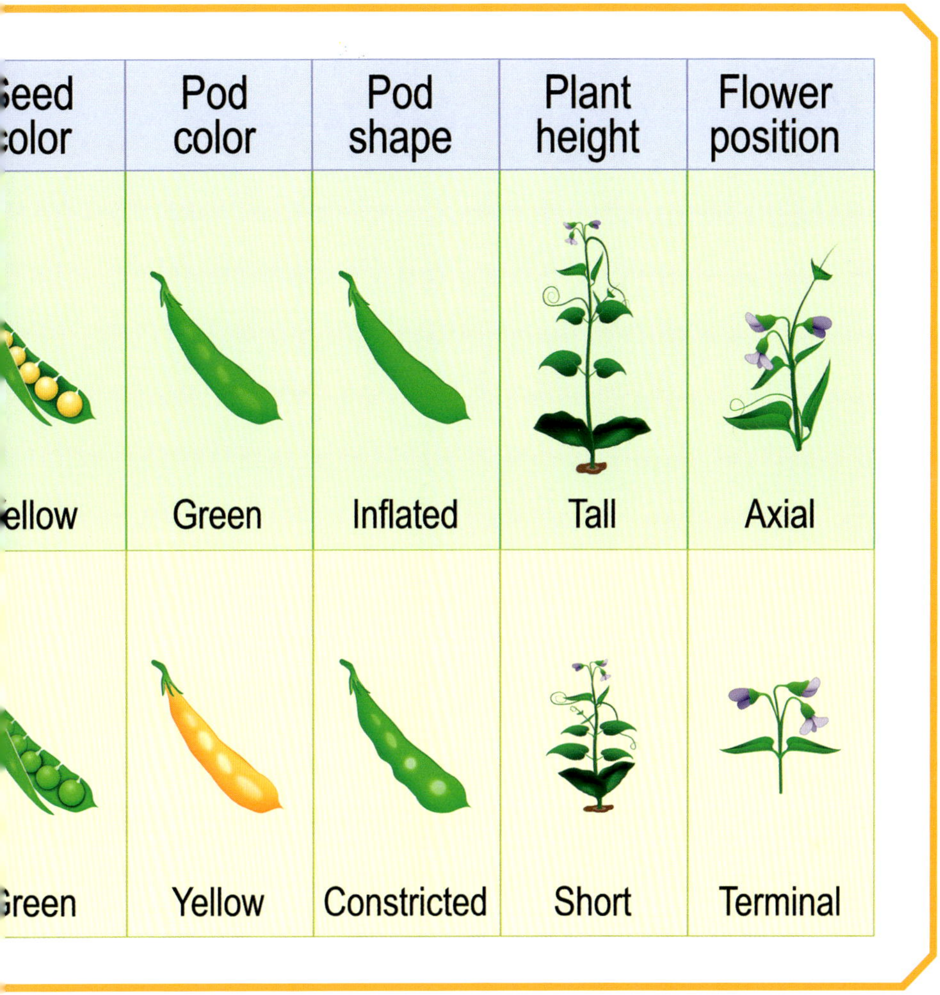

Mendel's experiments showed that some traits are dominant and appear more often. Other traits are recesive, and appear less often.

varieties were true-breeding by counting the offspring from every cross-pollination. Then he began cross-pollinating plants with specific traits. First, he tested one trait at a time. Later, he tested combinations of two and three traits. By carefully counting the offspring of thousands of crosses, Mendel collected a wealth of data.

CHAPTER 2

THE LAWS OF INHERITANCE

Gregor Mendel tested seven "either-or" traits in garden peas: seed color, seed shape, pod color, pod shape, flower color, flower position, and plant height. He meticulously observed the results from thousands of crosses, calculating ratios of traits in the offspring. From these results, he developed three laws of inheritance. Mendel deduced both the first and second laws from crosses testing one trait at a time. known as monohybrid crosses. More complex dihybrid and trihybrid crosses led to the third law. Mendel's crosses and his thought processes are complicated, and it helps to follow them step-by-step.

MONOHYBRID CROSSES

Mendel theorized that peas inherited two "factors," one from each parent, which determined a trait. He tested this hypothesis using monohybrid crosses. For example, he crossed a plant that bred true for round seeds with one that bred true for wrinkled seeds. These two plants were the mother and father in his cross. They were the first (parent) generation. This cross produced a second generation with all round seeds. The round-seed trait is dominant because it completely masks the presence of the wrinkled-seed trait. The masked form (wrinkled seeds) is recessive.

This diagram demonstrates Mendel's law of segregation and the 3:1 ratio. White flowers only appeared when two recessive alleles were paired—one-fourth of the time.

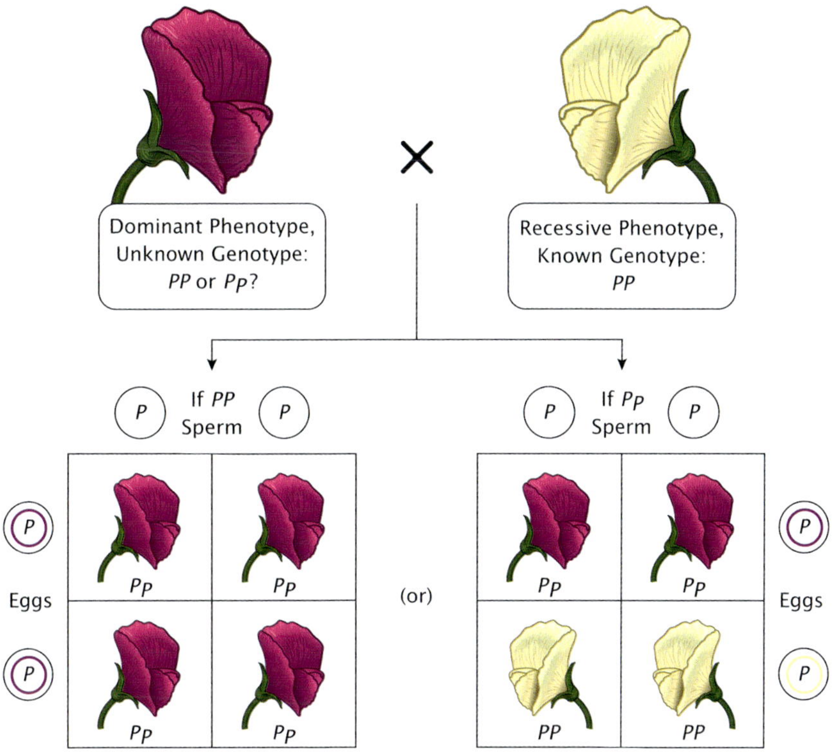

However, Mendel proved the recessive trait did not disappear by allowing members of the second generation to self-fertilize. This produced a third generation in which one-fourth of the seeds were wrinkled. What Mendel called factors are now called genes. The two forms of each gene—the "either-or" factors, such as round or wrinkled seeds—are called alleles of the gene.

Mendel knew he needed large sample sizes for accurate results. In his data, there are three times as many dominant plants as recessive plants. The larger the number of plants tested, the closer the ratio is to 3:1. This is because large sample sizes help even out random variations.

MENDEL'S CONCLUSIONS

For the third generation of pea plants, there were no blended forms. For example, crosses of yellow and green seeds did not produce yellowish-green or greenish-yellow seeds. The same held true for all traits tested. The fact that traits from both parents showed up unchanged in the third generation supported Mendel's hypothesis. This hypothesis became known as the theory of particulate inheritance, which says that traits are determined by factors (genes) that pass unchanged from generation to generation. This conclusion was a breakthrough in the study of inheritance. It disproved the old idea of "blending inheritance."

In Mendel's time, people did not know about cell structures that determined inheritance. However, Mendel was able to deduce what happened inside the cell. He reasoned that the two separate factors sorted randomly in each parent and entered the sex cells. In peas, the egg and pollen grain would each contribute one factor to determine a trait such as seed color or shape. If each second-generation parent produced equal numbers of each type of sex cell (for example, green-seed and yellow-seed cells), there was an equal chance of the offspring receiving either type. From this, Mendel deduced the law of segregation. This second law states that pairs of factors (alleles) segregate randomly and unchanged into separate sex cells.

Mendel's research led to many new discoveries in the field of genetics. Some scientists determined patterns of inheritance in other species and figured out more complex types of inheritance. Others studied cell structure, discovered chromosomes, and described cell reproduction and division. Still others tracked down the exact nature of Mendel's "factors"—the genetic material we now know as DNA.

THE LANGUAGE OF GENETICS

Mendel crossed-pollinated plants with visible traits, now called phenotypes. He counted the different phenotypes that appeared. From his data, he deduced the combination of alleles producing each phenotype. This combination is called the genotype. The genotype is hidden inside the cell's DNA.

Scientists developed a shorthand way to describe the alleles in genotypes. A capital letter stands for the dominant allele; for example, Y stands for yellow seeds. The lowercase y stands for the recessive allele of the same gene, such as green seeds. Each parent gives one allele—Y or y—to each offspring. So offspring have one allele from each parent to make a pair for each trait.

Some genotypes have two alleles of the same type. If both alleles are either dominant (YY) or recessive (yy), they're called homozygous genotypes. Heterozygous genotypes, on the other hand, have one recessive and one dominant allele (Yy).

THE PUNNETT SQUARE

The English geneticist Reginald Crandall Punnett (1875–1967) developed a simple diagram that can predict the outcomes of crosses. We call it the Punnett square. It illustrates the results of a cross by showing all possible genotype combinations. It demonstrates how recessive traits persist in a population. Furthermore, it also shows how different alleles segregate during sex cell formation and regroup in offspring.

This Punnett square shows a monohybrid cross of two heterozygous parents:

ALLELES IN EGGS / ALLELES IN SPERM	Y	y
Y	YY	Yy
y	Yy	yy

In 1905, Reginald Punnett wrote the first textbook about Gregor Mendel's work. It was titled *Mendelism*.

The Y and y across the top stand for yellow (dominant Y) and green (recessive y) alleles that the sperm in the pollen grain from a heterozygous plant can donate. The Y and y on the left side are possible alleles from the egg. The four squares inside the box show how the heterozygous parents' alleles can pair at fertilization. These are the only four possible combinations of these two alleles.

This Punnett square shows that a monohybrid cross of two heterozygous parents can produce three possible genotypes in the next generation. YY means the offspring receives a dominant allele from each parent and has a yellow phenotype. The sex cells of this offspring will all have the dominant Y allele. A Yy offspring has one dominant and one recessive allele. Since yellow is dominant over green, the green allele is masked and Yy genotypes also have yellow phenotypes. It makes no difference that the dominant Y allele came from the mother in one Yy offspring and from the father in the other. Half the sex cells produced by these plants will contain the Y allele. The other half will have the y allele.

These three offspring—YY, Yy, and Yy—represent the "3" in Mendel's 3:1 ratio. No matter how many plants he crossed, three out of every four offspring showed the dominant phenotype. One of the three had a homozygous dominant genotype, while the other two were heterozygous like the parent plants.

The third genotype (yy) features two recessive alleles. These offspring will be green because there are no dominant alleles present to mask the recessive allele. This genotype is the "1" in Mendel's 3:1 ratio.

CHAPTER 3

PREDICTABILITY AND PROBABILITY

Gregor Mendel thought alleles for a trait segregated randomly, or by chance, when they passed into sex cells, or gametes. If this were true, Mendel knew he could predict the results of crosses using probability. Probability is the chance, or likelihood, that a given event will occur when something is done repeatedly. Probabilities are expressed as fractions, decimals, or percents. For example, the probability of getting heads when flipping a coin is 50 percent; there's also a 50-percent chance of the coin landing on tails. So, the likelihood of heads landing up is a 1 in 2 chance, or 1:2.

In a monohybrid cross between two heterozygous parents, there's a one-in-four chance (1:4) that any offspring will show the recessive trait. This probability is expressed as 25 percent.

Mendel knew the parents in his initial crosses were true-breeding, but he didn't know the offspring's genotypes. He did testcrosses to determine this. A testcross is a cross between an individual and its homozygous recessive parent type. For example, Mendel might cross a yellow offspring (Yy) from the second generation with its green (yy) parent type. The offspring of such a testcross might have either a Yy or a yy genotype.

Flipping a coin can only have two outcomes: heads or tails. If you flipped a coin 100 times, it's likely that heads would land up 50 times, and tails would land up 50 times.

Punnett squares can predict the results of crossing an individual of each genotype with one that is homozygous recessive. If the genotype of the individual is homozygous dominant (YY), as shown in the left Punnit square in the chart on page 26, the offspring will all be yellow. If the genotype of the individual is heterozygous, or Yy, half the offspring will be yellow and half will be green. A researcher can determine the individual's genotype by comparing the actual results of the cross with the predictions shown in the Punnett square.

Mendel's testcrosses of second-generation offspring backcrossed with their green parent gave the needed information. These results showed the genotype of the second generation was heterozygous, not homozygous, because the green trait appeared. It also showed there were equal numbers of green and yellow alleles. If alleles had not sorted randomly (and equally), the seeds would not have been half yellow and half green.

YELLOW SEED
unknown genotype: either YY or Yy

	Y	Y
y	Yy	Yy
y	Yy	Yy

GREEN SEED
known genotype: yy

	Y	y
y	Yy	yy
y	Yy	yy

DIHYBRID AND TRIHYBRID CROSSES

Mendel also believed that "factors" for different traits were not connected to each other. He thought genes controlling different traits, such as seed color and seed shape, were independent and traveled separately when they entered gametes. To test this hypothesis, he conducted dihybrid crosses, using combinations of two genes.

The following Punnett square shows a dihybrid cross using two different pairs of alleles—round and wrinkled seeds, and short and tall plants. T represents tall (dominant) plant types and t represents short (recessive) plant types. R indicates round (dominant) seeds and r indicates wrinkled (recessive) seeds. Each parent donates one allele from each of the two pairs. So the possible gametes, shown outside the square, have two alleles, not one, and the offspring inside each box have two pairs of alleles.

		ALLELES IN SPERM			
		TR	Tr	tR	tr
ALLELES IN EGGS	TR	TTRR	TTRr	TrRR	TtRr
	Tr	TTRr	TTrr	TtRr	Ttrr
	tR	TtRr	TtRr	ttRR	ttRr
	tr	TtRr	Ttrr	ttRr	ttrr

A dihybrid cross results in 16 possible combinations of two gene pairs. Whenever a dominant allele is present, that allele appears in the phenotype. All four genotypes in the top row have at least one dominant allele for both height and seed shape. So all four will have tall, round phenotypes. In the second row, all four genotypes have at least one dominant allele for height, but two have the rr genotype, which will result in a wrinkled seed shape. So of these four, two will be "tall, round" and two will be "tall, wrinkled." Combining all similar phenotypes in this way gives a set of predictable mathematical ratios:

9/16 tall, round
3/16 tall, wrinkled
3/16 short, round
1/16 short, wrinkled

In shorthand form, this is a 9:3:3:1 ratio. Remember that a monohybrid cross has three different genotypes. For seed color, these were YY, Yy, and yy. A dihybrid cross has nine different possible genotypes. For plant height and seed shape, these were:

TTRR	TTRr	TTrr
TtRR	TtRr	Ttrr
ttRR	ttRr	ttrr

Mendel also experimented with trihybrid crosses, or crosses involving three traits. When he tested traits for height, seed shape, and seed color together, the same thing happened, only with more possible combinations. Three pairs of genes have 27 possible genotypes. The phenotype ratio for a trihybrid cross is 27:9:9:9:3:3:3:1. Twenty-seven of the offspring will have all dominant traits (tall height, round seed shape, yellow color). Only one of the offspring will have all recessive traits (short height, wrinkled seed shape, green color).

In Mendel's trihybrid crosses, every possible combination of traits appeared in the offspring in these predictable ratios. Inheritance patterns for each trait were the same, regardless if they were tested separately or together. This suggested that alleles for the different traits sorted and traveled randomly. They followed the laws of probability. This confirmed Mendel's hypothesis. From these findings, he developed his third law of inheritance—the law of independent assortment. It states that different pairs of genes undergo independent assortment during the formation of gametes. In other words, they sort randomly and travel separately into the egg or sperm cell.

This Punnett square shows another dihybrid cross using two different pairs of alleles—green and yellow seeds, and round and wrinkled seeds. You can see that the results are the same with a 9:3:3:1 ratio.

♂ Gametes

♀ Gametes	RY	Ry	ry	rY
RY	RR YY	RR Yy	Rr Yy	Rr YY
Ry	RR Yy	RR yy	Rr yy	Rr Yy
ry	Rr Yy	Rr yy	rr yy	rr Yy
rY	Rr YY	Rr Yy	rr Yy	rr YY

MENDEL'S LAWS OF INHERITANCE

In many cases, only two of Mendel's ideas are given the status of laws or principles: segregation and independent assortment. Particulate inheritance is not always included, but without this breakthrough, the other two laws would not have been possible.

1. The theory of particulate inheritance—Discrete factors (genes) are inherited unchanged from generation to generation.

2. The law of segregation—Factors (genes) controlling a single trait pass separately and independently into different gametes.

3. The law of independent assortment—Alleles controlling different traits sort independently into different gametes.

MENDEL'S GIFT TO THE WORLD

Mendel chose to study exactly the right garden pea traits. by luck or on purpose. He wasn't aware of the existence of chromosomes or that many genes are located together on the same chromosome. The traits Mendel chose happened to be either on different chromosomes or far enough apart so that they sorted independently. This led him to the very neat conclusion that all traits travel separately. Later geneticists found things are not always so simple. Does this mean that Mendel was wrong?

Not at all. His laws are absolutely correct for the traits he studied and for all situations where two genes are located on different chromosomes. But other rules apply when genes for two traits are located near each other on the same chromosome. After eight years of studying the genetic inheritance of garden peas, Gregor Mendel had set the stage for the development of the entire field of genetics. His laws of inheritance formed the basis for all the research that followed. Mendel died in 1884, before people realized the importance of his contributions. Before his death, he wrote, " My scientific studies have afforded me great gratification; and I am convinced that it will not be long before the whole world acknowledges the results of my work." He was right!

Shown here is the garden at the abbey in Brno where Mendel conducted his experiments.

CHAPTER 4

COMPLEX INHERITANCE PATTERNS

Once Gregor Mendel's experiments became known, other geneticists began to study other traits and species. They soon realized that not all traits were inherited as neatly as Mendel's laws suggested. This is because not all traits are clearly dominant or recessive. Others appeared to have more than two forms, or alleles. Not all plants have just two colors of flowers. Some can come in three or more color varieties. Humans have four blood types. Scientists discovered that some traits appeared much more frequently in men than in women. And some traits appeared to travel together. If one trait appeared in an offspring, another (often totally different) trait also appeared. These clues suggested not all inheritance followed Mendel's laws.

INCOMPLETE DOMINANCE

The self-crossing heterozygous purple flowers of Mendel's garden peas produced a 3:1 ratio of purple to white flowers in the next generation. That means the trait of purple flowers is completely dominant over that of white flowers. But some other crosses produce a different result. When a pink snapdragon is self-crossed, for example, half the flowers are pink. One-fourth of the offspring are homozygous (RR) and red, one-half are heterozygous (Rr) and pink, and the other one-fourth are homozygous recessive (rr) and white.

This situation is called incomplete dominance. Both red and white alleles show some dominance, but neither color can completely mask the presence of the other. If Mendel had studied snapdragons instead of garden peas, he might have thought the Greek idea of blending inheritance was correct.

Gardeners love snapdragons because of their beautiful colors and because they return each spring—often in different colors!

ENVIRONMENTAL INFLUENCE

Genes do not act alone to control traits. The environment always plays an important role. Melanin production in some animals is related to temperature—the points on a Siamese cat are darker than the torso because the body temperature of the ears, tail, and feet is lower. Environmental differences affect all genes. Even identical twins are unique because their environments are always slightly different.

CODOMINANCE AND MULTIPLE ALLELES

In codominance, two or more dominant alleles are fully expressed in different parts of the organism. Expression of the alleles is both independent and equal. A good example of codominance is the rhododendron. They have alleles for red and white petal color, named PR and PW. P stands for petal

RHODODENDRON BLOSSOM DISPLAYING CODOMINANCE

color. The capital letters R and W indicate that both colors are dominant. When a red rhododendron (PRPR) is mixed with a white one (PWPW), the resulting offspring have the genotype PRPW. They have patches of white and red. The blossoms aren't pink, as in incomplete dominance. The rhododendron displays both colors, so blossoms look mottled or speckled.

The system of ABO blood types in humans illustrates codominance, but it also illustrates another complex inheritance pattern: multiple alleles. In the human species, three different alleles determine ABO blood types, and each individual receives only two of these, one from each parent. Two of the three alleles, called IA and IB, are codominant. The third allele, i, is recessive. The dominant alleles stimulate the buildup of proteins, either antigen A or antigen B, on the surfaces of red blood cells. The recessive allele does not produce an antigen.

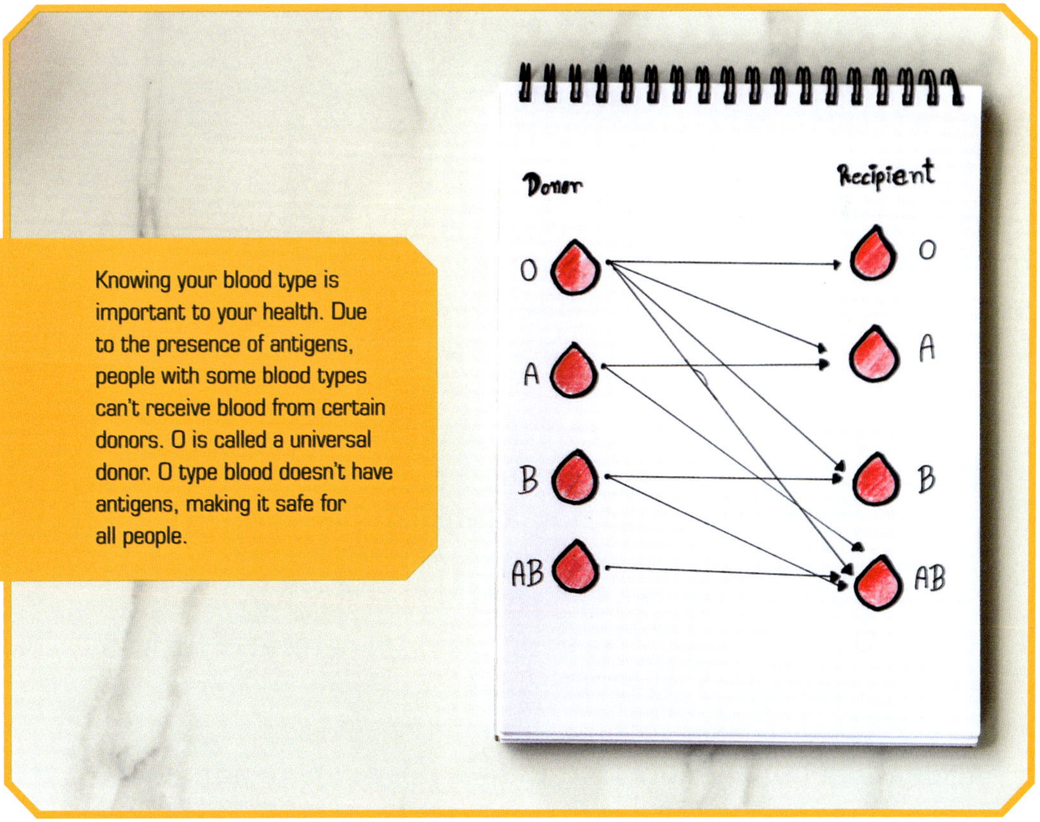

Knowing your blood type is important to your health. Due to the presence of antigens, people with some blood types can't receive blood from certain donors. O is called a universal donor. O type blood doesn't have antigens, making it safe for all people.

These three alleles combine in pairs give six possible genotypes. They can produce four different phenotypes—blood groups A, B, AB, and O. The table shows the combinations of genotypes and the phenotypes they produce.

BLOOD TYPE (PHENOTYPE)	POSSIBLE GENOTYPE
A	I^AI^A or I^Ai
B	I^BI^B or I^Bi
AB	BI^AI^B
O	ii

This chart shows that people with type A or type B blood can be homozygous for the dominant allele. They also can be heterozygous, having one dominant and one recessive allele. People with type AB blood have codominant alleles. People with type O blood have two recessive alleles.

Blood types illustrate something important about how dominance truly works. You may think dominant alleles outnumber recessive alleles in a population, but this is not always the case. For example, the most common blood type in the human population is type O, which has two recessive alleles. Forty-five percent of Caucasians, 51 percent of African Americans, and 57 percent of Hispanics are type O. This shows that dominance just describes the ability to mask the recessive allele when both are present. It does not relate to numbers.

POLYGENIC INHERITANCE

Many traits in a population show continuous variation. That means there's a range of variation, but no precise groups, such as purple versus white color in garden pea plants. Think about human traits such as hair, eye, and skin color. These continuous traits result from polygenic inheritance. "Polygenic" means they're controlled by two or more gene pairs, instead of one. Hair, eye, and skin color result from a dark pigment called melanin. Two genes (A and B) control melanin production for skin color. Melanin production is dominant. Lack of production is recessive. A person who is homozygous recessive

Many animal species exhibit cases of albinism.

for both genes (aabb) produces no pigment. This results in a condition called albinism, where a person has very light skin, hair, and eye color. A person with only one dominant allele (Aabb or aaBb) is very light-skinned. A person with two dominant alleles (AAbb, AaBb, or aaBB) has medium skin color. A person with three dominant alleles (AABb or AaBB) has very dark skin, and one with all dominant alleles (AABB) has completely black skin.

Melanin also colors the iris of the eye. The more melanin produced, the darker the eye color. One main gene controls eye color. Dominant allele B controls production of brown, green, hazel, or gray eyes. Recessive allele b controls production of blue eyes. A blue-eyed person has very little melanin. However, several other genes control the amount and distribution of melanin, so there's a continuous range of eye colors, as there's a continuous range of skin colors. The same thing can be said of human hair.

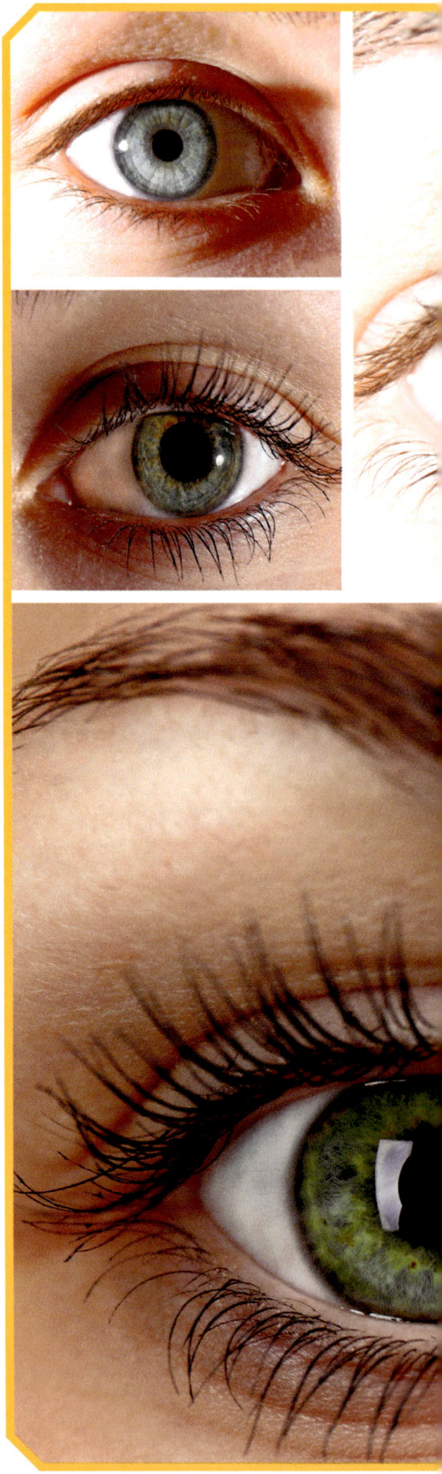

Brown eye color is a dominant trait, and the most common eye color in the world.

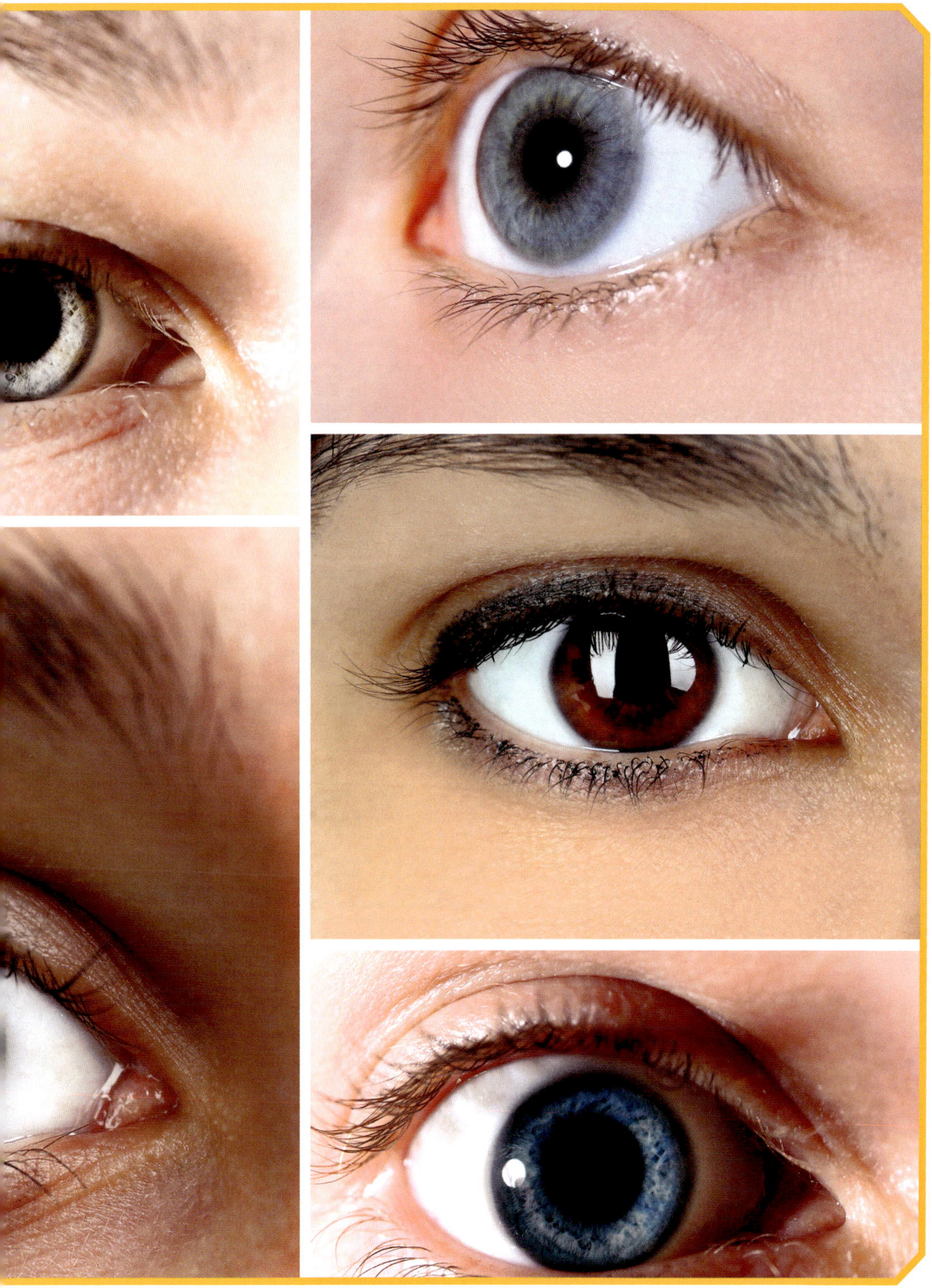

CHAPTER 5

THE PROCESS OF MITOSIS

There are two kinds of cellular reproduction: mitosis and meiosis. The first kind, mitosis, makes exact copies of a cell's DNA and creates a new cell with its own nucleus. Mitosis (followed by the division of the rest of the cell) keeps the number of body cells constant. It replaces dead and injured cells and allows a body to grow. Every person starts out life as a fertilized egg, or zygote. An adult human body contains about 75 trillion to 100 trillion cells, all of which came from that original zygote, and all were through the process of mitosis.

This diagram shows the steps of mitosis.

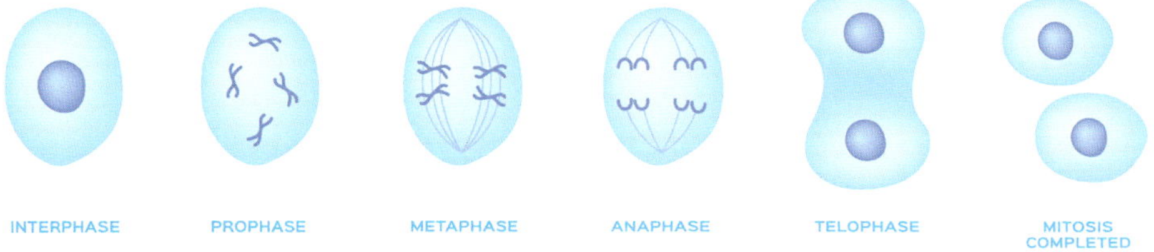

GENETICS AND CELL REPRODUCTION

Mendel's laws explain how he believed traits were passed from one generation to the next. But Mendel had no direct evidence because he didn't know what really went on inside the cells of his peas. The explanation came with the discovery of chromosomes during the late 19th and early 20th centuries. People thought genetic information was contained in the cell nucleus, but they didn't know the nature of that information or how it was transferred. A German anatomist named Walther Flemming used a microscope to investigate cell nuclei. He stained cells with dyes, which produced a very dark network of fibers inside the nucleus. Flemming called it chromatin. He saw that during nuclear division (mitosis), the network became a series of threadlike bodies. These bodies were eventually named chromosomes.

Scientists believed that genetic material was stored in cell nuclei, but could not prove it. As microscope technology improved, they discovered chromosomes and the genes that they contain.

INSIDE DNA

Mitosis is the process that transfers genetic information from cell to cell. This begins with the genetic material, or DNA molecule. "DNA" is short for "deoxyribonucleic acid." Most DNA in complex organisms resides in the cell nucleus. Each chromosome in the nucleus is made up of genes. Each gene controls or helps control a single trait. Together, all the genes control all the traits that make up the body's structures and functions. These include everything from hair color to hormone production.

This diagram shows how a chromosome is made up of many genes.

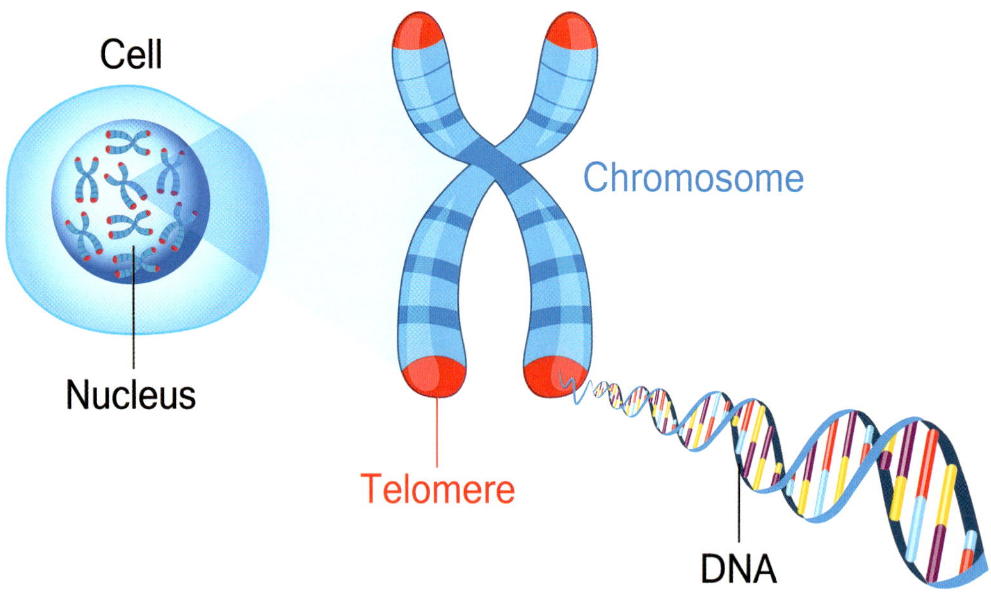

Each organism has a specific number of chromosomes. Humans usually have 46 chromosomes in 23 pairs. Mendel's garden peas have 14 chromosomes in seven pairs. This is the diploid, or 2n number, of chromosomes.

When a cell is ready to divide, it grows to a certain size. Then it makes a copy of its DNA. The two DNA copies separate, and the cell divides to form two new cells. Each new cell now has one complete copy of the DNA. The two new cells are called daughter cells and are identical to the original cell, called the mother cell. This sequence repeats, becoming a cell cycle.

THE PHASES

Mitosis is divided into steps, or phases. The phases are somewhat difficult to define because the cycle is a continuous process. However, you can identify each phase of dividing cells using a microscope slide of an onion root tip. The cell spends up to 90 percent of its time in interphase before mitosis begins. Individual chromosomes aren't visible during this time. They appear as the darkly stained network described by Flemming. During interphase, the cell increases in size and its DNA replicates.

Once the cell contains two copies of all its DNA, mitosis begins. During the first stage, prophase, DNA molecules shorten and coil into tight, visible chromosomes that spread randomly through the cell. They are double structures with two parts, or chromatids, held together by a centromere. Each chromatid contains one copy of the DNA. During prophase, the nuclear membrane begins to break down and two structures called centrosomes drift to opposite sides of the cell.

The second stage is called metaphase. The centromeres of all the chromosomes line up along the center, or equator region of the cell. The chromatids hang to either side of the centerline. Each chromosome moves independently, so they can line up in any order. Members of a pair are not necessarily near each other in line. Spindle fibers stretch between each centromere and the two centrioles, which are located at opposite sides of the cell.

During anaphase, centromeres break apart. The chromatids separate and travel to opposite sides of the cell. Once they separate, the chromatids are called chromosomes. In animal cells, the spindle fibers attached to each centromere pull them away from the middle of the cell. Next, during telophase, a new nuclear membrane forms around each new set of DNA. The chromosomes relax and return to their interphase state. Near the end of telophase, during the process called cytokinesis, the whole cell divides and a new cell membrane forms around each set of DNA.

The end result is two new daughter cells. Each new cell has one complete copy of the body's DNA, containing all the genes. The cell still has the diploid number of chromosomes—generally 46 in humans. The two new, genetically identical cells, now in interphase, will grow in size. Soon they will duplicate their DNA, and each may continue its own cell cycle.

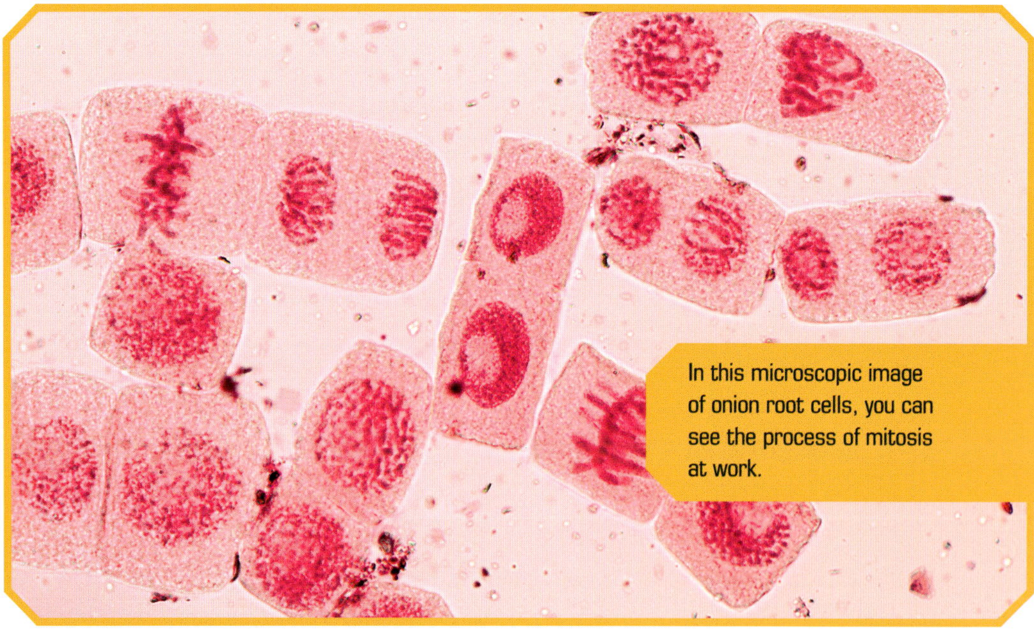

In this microscopic image of onion root cells, you can see the process of mitosis at work.

MITOSIS OUT OF CONTROL

Cell division is a remarkably well-regulated biological process. An adult human has trillions of cells! Many have very specific jobs and do those jobs very well. Sometimes, however, mitosis goes wild. A mutation causes the cell cycle to break down and a cell to become a cancer cell. It divides without control and forms a lumpy mass of cells called a tumor.

There are many kinds of cancer, but all cancer cells show abnormal, disorganized growth. They crowd out healthy cells and prevent them from doing their jobs. Cancer cells don't carry out the functions of normal cells. They break apart, move to other locations in the body, and form new tumors. This is a process called metastasis. Cancer cells may cause death if untreated. Because the cells are unable to control their own growth, doctors must kill the cells through a variety of treatment. These include surgery, drugs, radiation, or other treatments.

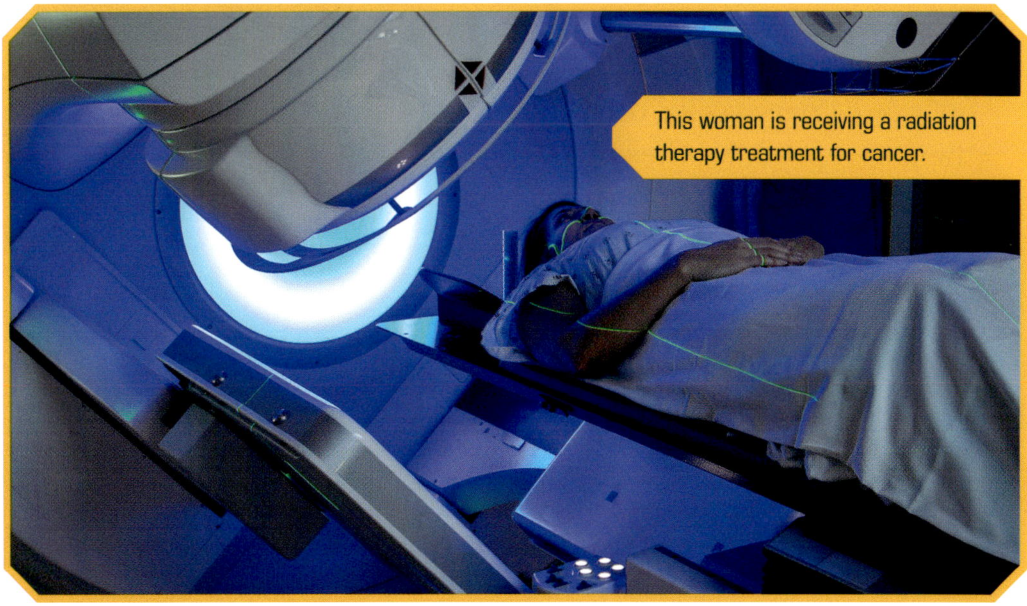

This woman is receiving a radiation therapy treatment for cancer.

RATE OF CELL DIVISION

The rate at which cells reproduce depends on the type of cell. Skin cells, for example, are very active cells. The whole cycle takes about 24 hours, but mitosis itself lasts only about an hour. Under normal conditions, the entire epidermis, or outer layer of skin cells, replaces itself by mitosis every 52 to 75 days. When an injury occurs, mitosis speeds up to quickly replace the damaged cells.

Other cell types replace themselves very slowly. Until recently, scientists thought nerve cells did not replace themselves. They thought each person was born with all their nerve cells, and that the brain slowly degenerated over a lifetime. However, we now know that some nerve cells do slowly replace themselves. If scientists can "turn on" young nerve cells, or stimulate them to reproduce, these new cells can replace missing or damaged cells. Eventually, this could help treat diseases, such as multiple sclerosis and Parkinson's disease, and it could help repair injuries to the brain and spinal cord.

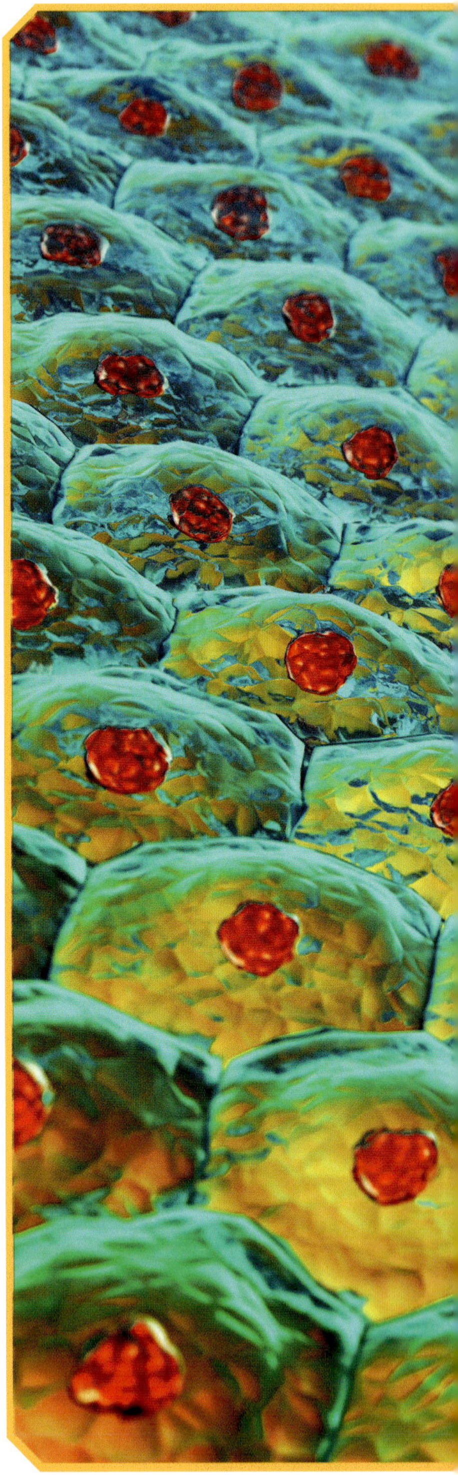

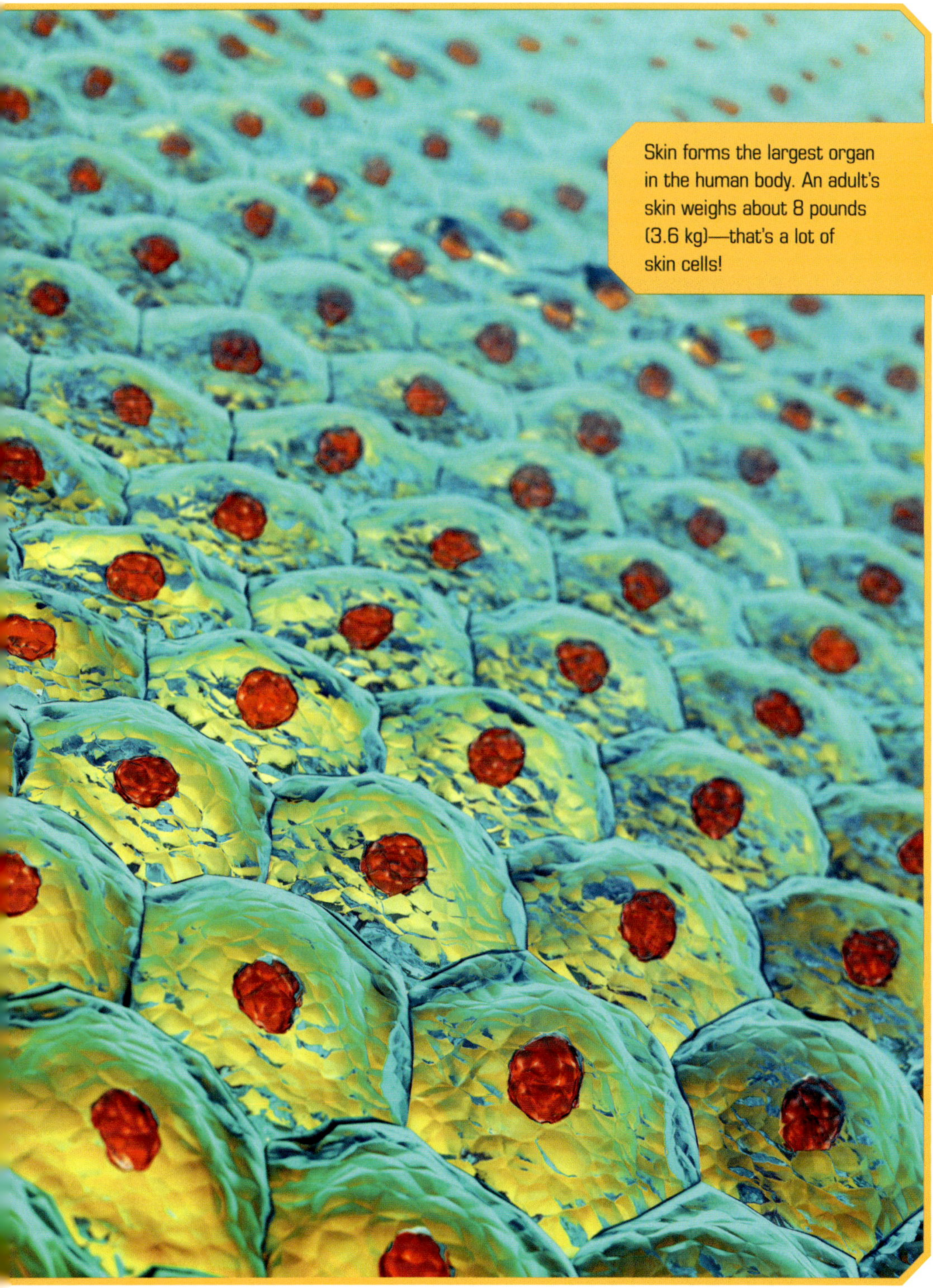

Skin forms the largest organ in the human body. An adult's skin weighs about 8 pounds (3.6 kg)—that's a lot of skin cells!

CHAPTER 6

THE PROCESS OF MEIOSIS

The second kind of nuclear reproduction, meiosis, creates gametes. These are the sex cells that pass from one generation to the next. Sexual reproduction begins with gamete production. Female gametes, or eggs, are made in the ovaries. Males gametes, or sperm, are made in the testes. Sexual reproduction is the major cause of variation. It's the reason why no two organisms are alike.

A few years after Walther Flemming's work on mitosis, another German scientist, Theodor Boveri, studied chromosomes in gametes of *Ascaris*, a parasitic roundworm with only two pairs of chromosomes. During gamete formation, Boveri observed the chromosome number decreased from four to two. The number returned to four after fertilization. This was one of the first descriptions of meiosis.

At the turn of the 20th century, American geneticist and physician Walter Sutton studied meiosis in the lubber grasshopper. He identified its 11 pairs of chromosomes plus a single sex chromosome. He followed the progress of each chromosome and described the stages of meiosis. Perhaps most significant, Sutton connected his observations with Mendel's paper, which was published 36 years before. He argued that

Mendel's "factors" moved into sex cells by meiosis. Sutton's 1902 paper was the first statement of the chromosomal theory of inheritance.

Sutton grew up in Utica, New York, where the lubber grasshopper was a common insect.

In 1903, Sutton made another major breakthrough. He knew that half the chromosomes had come from the grasshopper's father and half had come from its mother. However, they didn't line up on "mother" and "father" sides during gamete formation. The chromosome pairs lined up randomly down the cell center before dividing. When they separated into new cells, different gametes received different combinations of parental traits. Through his experiments, Walter Sutton was the first person to explain how the formation of sex cells results in variation. This was a question Darwin and Mendel had wondered about, but they never arrived at an answer.

MEIOSIS AND VARIATION

Sutton showed that sexual reproduction, which requires a type of cell division different from mitosis, causes variation, Gametes produced during meiosis have half as many chromosomes as cells produced during mitosis. That means they have only half as much genetic information. Also, genes are often rearranged during meiosis. These gametes have slightly different gene combinations than the person who made them. Variation also occurs during the process of fertilization. This is when an egg from a female and a sperm cell from a male fuse to form a zygote. Each parent has slightly different alleles. So fertilization gives the offspring a new mixture of alleles—half from the mother and half from the father. It has some characteristics of both parents, but it's also different from both parents.

MEIOSIS I

Meiosis occurs in two parts called meiosis I and meiosis II. The key difference from mitosis occurs during meiosis I when the chromosome number decreases by half. This is called reduction division. A gamete begins as a regular body cell in the ovary or testis with the diploid number of chromosomes. During interphase, this cell duplicates its DNA and makes doubled chromosomes, with two chromatids each—just as in mitosis.

This diagram shows the process of meiosis I.

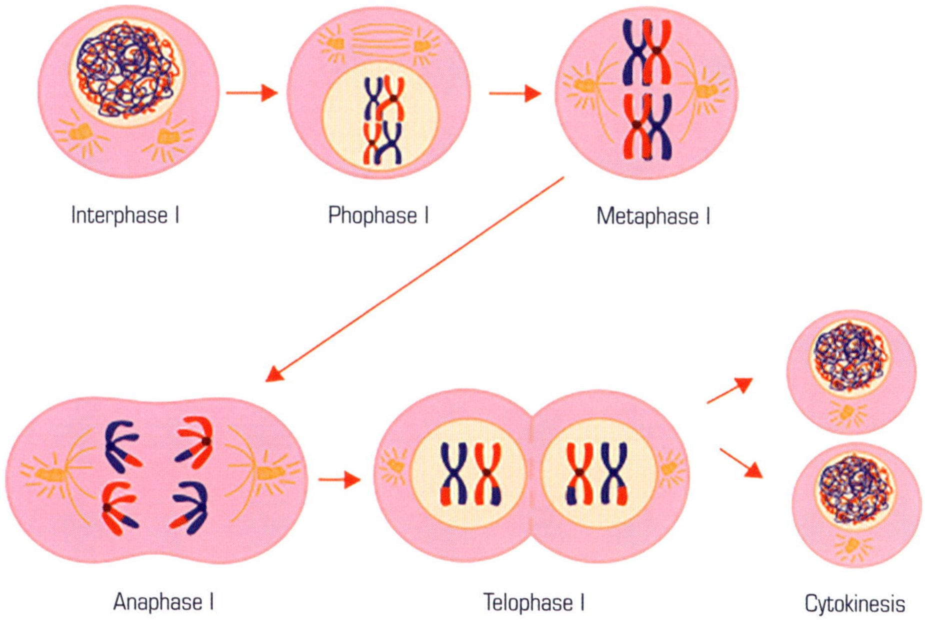

In prophase of meiosis I (or prophase I), the nuclear membrane breaks apart. The DNA shortens and coils to form visible chromosomes, as in mitosis. The chromosomes are randomly mixed, and the centrioles begin to migrate to opposite poles of the cell. However, metaphase I is different from mitosis. During meiosis, chromosomes don't line up in single file down the center of the cell. Instead of 46 individual chromosomes, they form 23 pairs. Members of homologous pairs find each other and line up side by side, one on either side of the cell's centerline. Spindle fibers stretch from each centromere to the centrioles at the two poles. By the end of metaphase, the chromosomes are neatly arranged in pairs, rather than lined up single file.

When homologous chromosomes pair up during metaphase, crossing-over can occur. During this process, the two inside chromatids get tangled together and end up exchanging genes. This changes the assortment of alleles on the homologous chromosomes. Some of the original mother's genes are now on the original father's chromosome, and vice versa. When the chromosomes separate, each gamete has a new combination of alleles. Crossing-over is one way meiosis creates variation in gametes.

During anaphase I, the paired chromosomes—some now with different gene combinations—move apart and drift toward opposite sides of the cell. Homologous pairs have separated forever. This is the second way variation occurs during meiosis. Nuclear membranes form around the two new sets of chromosomes during telophase I. Chromosomes return to their relaxed interphase state and cytokinesis occurs. At the end of meiosis I, there are two new daughter cells. Each cell has only half the number in the original cell. This is the haploid, or n, number. Each gamete will give only half the parent's genes to an offspring.

MEIOSIS II AND GAMETOGENESIS

Both daughter cells now carry out a second round of division, which is known as meiosis II. The nuclear material divides by exactly the same process as mitosis this time. Each new cell gets one chromosome (consisting of a single chromatid with one copy of the DNA) for each homologous pair in the original cell. Meiosis II begins with two cells. The end result is four cells. Because reduction division and crossing-over in meiosis I create variation, all cells are different from the original mother cell.

This diagram shows the steps of meiosis II.

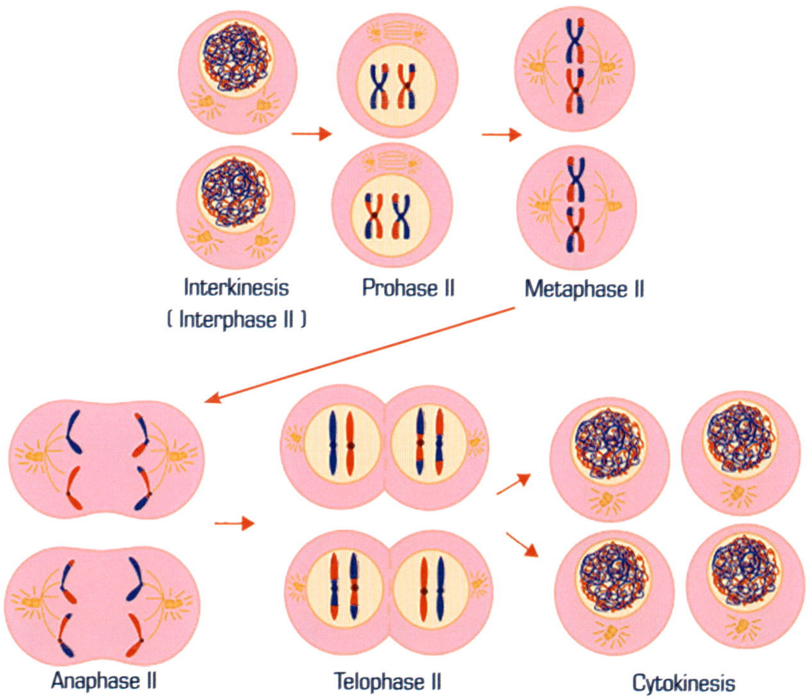

Gametogenesis is the formation of gametes. But this isn't complete when chromosomes separate. The process is different in human males and human females.

In human males, the process of cytokinesis during meiosis II results in equal division of the cytoplasm and it ends with four same-sized cells. These four cells begin to change shape. Most of the cytoplasm forms into a long tail. Gametogenesis in males makes four tiny but functional sperm cells. All the father's genetic material is preserved, although not all of it will be used because only one sperm can fertilize an egg. Nevertheless, the genetic material is all available for fertilization.

Nuclear division in females happens exactly the same as in males, but the division of cytoplasm is unequal. One daughter cell in meiosis I receives most of the cytoplasm, so it's bigger than the other. This happens again during meiosis II. The largest cell becomes an egg cell. The other three cells, called polar bodies, die. So any mother cell only forms one functional egg cell, and one-half the mother's genetic information is lost.

FERTILIZATION

Genes are thoroughly mixed during meiosis. Crossing-over occurs in metaphase I. Paired chromosomes separate randomly in anaphase I. Genetic information is lost when an egg is formed. So the eggs and sperm produced are considerably different from the parent's starting cell. This is how meiosis and gametogenesis produce variations.

Fertilization causes even more variation. Because of reduction division, human egg and sperm cells usually have only 23 chromosomes each—one member of each pair, randomly selected from the original pair. Only one in millions of sperm cells will fertilize the egg cell. The genetic mix of this one lucky sperm cell will be entirely random. In humans, only one egg cell per month will mature and be available for fertilization. Again, the genetic mix in the egg cell will be random.

When the single set of genes in the egg combines with the single set in the sperm, the third and final instance of variation occurs. The mother's and the father's genes will mix and form an entirely new combination of genes. The result will be a unique individual.

MEIOSIS OUT OF CONTROL

Like mitosis, meiosis can go wrong. When chromosome pairs line up during meiosis I, members of a chromosome pair sometimes stick together. Two chromosomes will go into the same cell, and neither will go into the other cell in that division. The resulting gametes will either have an extra chromosome or they will lack one. Failure of chromosomes to separate correctly during meiosis is called nondisjunction. It results in specific genetic disorders.

Chromosomes contain many genes. An extra or missing chromosome affects the entire body. An extra copy of chromosome 21 causes trisomy 21, commonly called Down syndrome. People with Down syndrome often have a number of easily recognized physical traits. These people may also have cognitive disabilities and heart defects.

Trisomy 18 is lesser known genetic condition called Edward syndrome. People with this condition have a extra copy of chromosome 18. Symptoms include severe intellectual disability, low birth weight, physical abnormalities, and congenital heart defects, among others. Many people with trisomy 18 don't live to be teenagers.

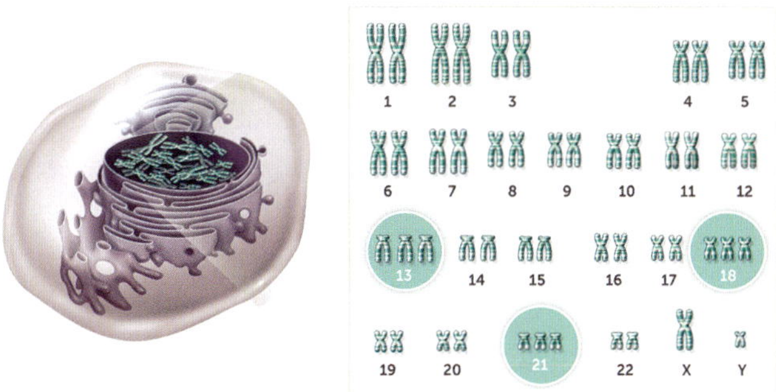

Trisomy 13, also known as Patau syndrome, results in numerous mental and physical problems. Most babies born with this condition don't live very long, but others live for many years.

CHAPTER 7

GENE LINKAGE

During meiosis, linked genes travel together. Linkage occurs because chromosomes, not genes, separate during meiosis. When gametes are formed, each separated chromosome passes into a particular gamete. All the genes on that chromosome go along for the ride, like cars on the same train. So, dihybrid crosses involving linked traits don't show the same 9:3:3:1 ratios found in traits that assort separately. They don't follow Mendel's third law and undergo independent assortment.

Gene linkage has the opposite effect of meiosis and fertilization, which increase variation. Linkage decreases variation. When a cross produces more parental-type offspring than predicted by independent assortment, this is a clue that linkage is occurring.

The simplest way to define linkage groups, or genes that travel together, is by the number of chromosome pairs in the cells of an organism. Most humans have 23 pairs of chromosomes and 23 linkage groups. Mendel's pea plants have seven pairs of chromosomes and seven linkage groups. Linkage, however, isn't that simple. Because of crossing-over, genes located on the same chromosome can still behave as though they are not linked.

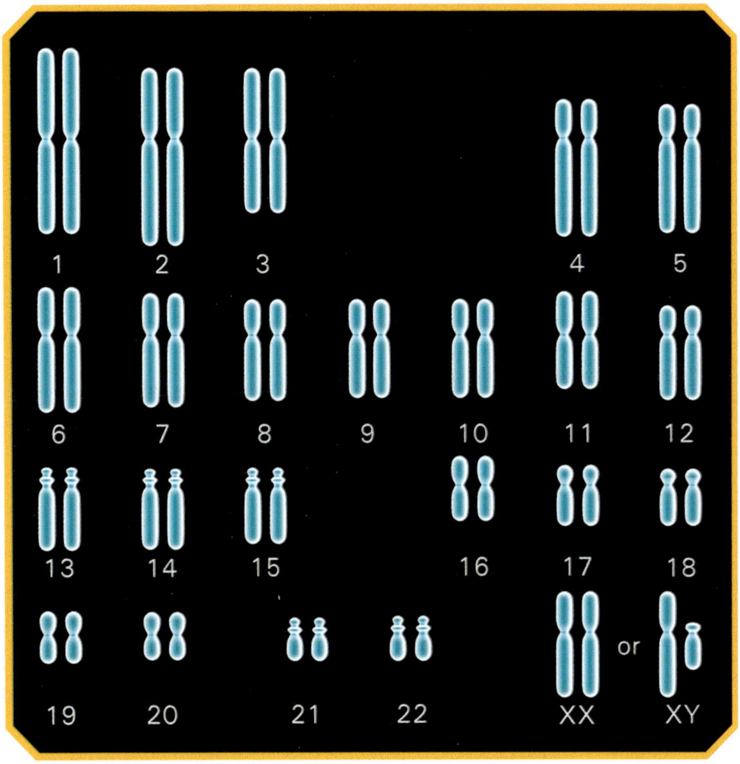

A karyotype is a visual representation of the number and appearance of chromosomes in a species' cells. Shown here is the human karyotype. Note that chromosome 23 is what determines sex traits in humans.

DNA STRUCTURE

To understand how linkage occurs, it necessary to know something about chromosomes and the structure of DNA. A chromosome is made up of a very long molecule of DNA associated with numerous protein molecules. This DNA molecule is made of tiny repeating units called nucleotides. Each nucleotide contains an even tinier unit called a base. There are only four types of bases in the DNA molecule. The DNA molecule consists of two strands in a shape called a double helix. The bases connect in specific pairs to hook the strands together.

The sequence of base pairs along the strand defines different genes. Genes vary greatly in length. Chromosome 1 is the longest and is made up of about 249 millions base pairs. Each gene contains a code determined by the sequences of bases along its length. This genetic code tells the cell how to make a certain protein. Together, the body's proteins determine the traits and behaviors that make each person unique. Environmental factors affect how traits are expressed. as we have seen. Also, most traits result from more than one gene. But every protein must be made correctly and in a precise amount for the trait to be expressed.

People have about 20,000 identified genes scattered along their chromosomes. Each gene helps make as many as six different proteins. Between the genes are long, repeating segments of noncoding (nongene) DNA, often misleadingly referred to as "junk DNA." Altogether, humans' 46 chromosomes have more than 3 billion base pairs. Probably only about 5 percent of these are actually part of a gene.

GETTING LINKED UP

The distance between two genes is measured in base pairs (bp), which determine how likely two genes are to be linked. If two genes on a chromosome are closer than 50 million bp, they're likely to be linked. This follows Mendel's law of independent assortment. However, if they're farther apart than 50 million bp, there's an increased chance they won't travel together. This is because of crossing-over.

A locus is a specific location of a gene along a chromosome. If two gene loci are close together, crossing-over will probably not cause the genes to be separated. But the farther apart the two loci are, the more likely it is that breaks and crossing-over will occur between them. If two genes are far enough apart, they act like they're on different chromosomes.

MORGAN'S FRUIT FLIES

In 1933, American geneticist Thomas Hunt Morgan won the Nobel Prize in Physiology or Medicine for discovering and describing patterns of gene linkage. He proved that genes are linked on chromosomes and are responsible for hereditary traits. After studying *Drosophila*—a genus of flies commonly known as fruit flies—for many years, he was the first scientist to identify and explain the crossing-over phenomenon.

Six Nobel Prizes have been awarded to scientists who've made genetic discoveries while studying flies of the *Drosophila* genus. The species shown here is *D. melaogaster.*

Morgan and his students knew that certain species of fruit flies have four pairs of chromosomes, including one pair of sex chromosomes. As in humans, female flies have two large X chromosomes and males have one X and one much smaller Y chromosome. However, unlike the situation in humans, the fruit fly's Y chromosome does not determine its maleness. A fruit fly is male if only one X chromosome is present and female if there are two. With only one X chromosome, there's never a dominant chromosome to mask the recessive allele. Morgan and his students discovered 40 different *Drosophila* mutations over the course of several decades. They bred large numbers of each mutant type. Then, following Mendel's lead, they did testcrosses to determine patterns of inheritance.

One mutation Morgan studied was eye color. After a year of testing, Morgan noticed a white-eyed male, which was a new mutation. Testcrosses with the white-eyed mutation led Morgan to question Mendel's laws. When he crossed red-eyed females with white-eyed males, first- and second-generation offspring both followed Mendel's laws. First-generation flies were all red-eyed and the second generation showed a 3:1 ratio of red to white. However, when he crossed white-eyed females with red-eyed males, both the first and second generations showed half red-eyed and half white-eyed flies.

Morgan reasoned that the gene for eye color must be located on the X chromosome. The white-eyed allele was recessive, so it always showed up in males that possessed the allele because they have only one X chromosome (rO). The Y chromosome had nothing to do with it. The allele had an effect in females only when it was present as a double recessive (rr). Morgan did not just assume his assumptions were correct. He and his students counted eye color in 13 million flies before they were convinced!

LINKAGE ANALYSIS

A linkage analysis determines if gene loci on a chromosome are close enough to be considered linked. It shows how often crossing-over occurs between two or more genes. To do a linkage analysis, scientists once again follow the lead of Gregor Mendel: they do testcrosses between heterozygotes and recessive homozygotes and count the phenotypes of the offspring. If there's no linkage, all combinations of phenotypes will appear in approximately equal frequencies. However, the proportions will be unequal if linkage is occurring. The table shows a 20 percent cross-over, with the two rarer phenotypes each being produced 10 percent of the time.

Linkage in a Dihybrid Testcross (AaBb x aabb)	
GENOTYPE	**PROPORTION**
Aabb	40 percent
aaBb	40 percent
AaBb	10 percent
aabb	10 percent

The map distance is the amount of cross-over between genes at two loci. It is calculated by dividing the total number of recombinant (cross-over) offspring by the total number of offspring. The result, multiplied by 100, gives the percent of recombinants. In humans, one map unit on a chromosome represents about 1 million base pairs and 1 percent recombination.

SEX-LINKED TRAITS

Sex-linked traits are traits located on a sex chromosome—the other chromosomes are called autosomes. The human X chromosome has 1,141 genes and is more than 150 million base pairs long. The Y chromosome is tiny in comparison—it has only 255 genes and about 50 million bp. One Y-chromosome gene, the SRY gene (for sex-determining region Y), causes testes to form during early development. They begin making the hormone testosterone, and the baby becomes what is physically considered a boy. A baby with two X chromosomes and no Y develops ovaries and becomes what is physically considered a girl.

Thomas Hunt Morgan and his students did the first linkage analyses. They studied linkage in 40 *Drosophila* mutations, located them on specific chromosomes, and determined which were linked and which were not. One of Morgan's students, Alfred Sturtevant, made the first gene map while an undergraduate. It shows relative gene locations on the fruit fly's X chromosome. Morgan's group later mapped gene locations on the fly's other three chromosomes.

This illustration, which shows the basic process of crossing-over in chromosomes, appeared in Morgan's 1916 book *A Critique of the Theory of Evolution*.

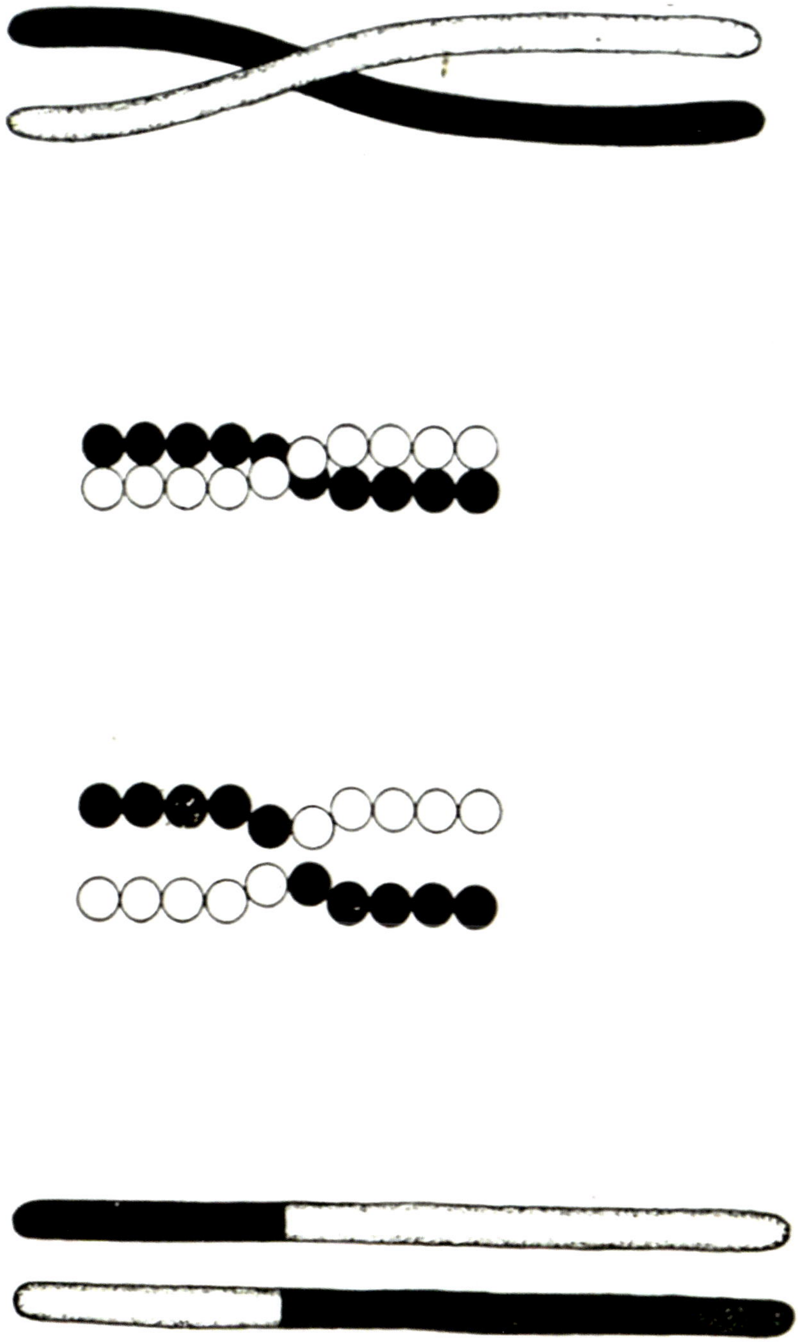

CHAPTER 8

GENE SEQUENCING AND MAPPING

The mid-20th century was a booming time for genetics research and discovery. In 1953, American James Watson and Englishman Francis Crick described DNA as a double helix. In 1966, the "genetic code" was cracked when research showed that each group of three bases along a gene coded for an amino acid. Pulling amino acids together in the same sequence as DNA's base triplets made a protein. This proved that DNA's function was to make proteins.

Genetic code is represented by long strands of letters. Each letter stands for a nucleotide base segment in a chain of DNA.

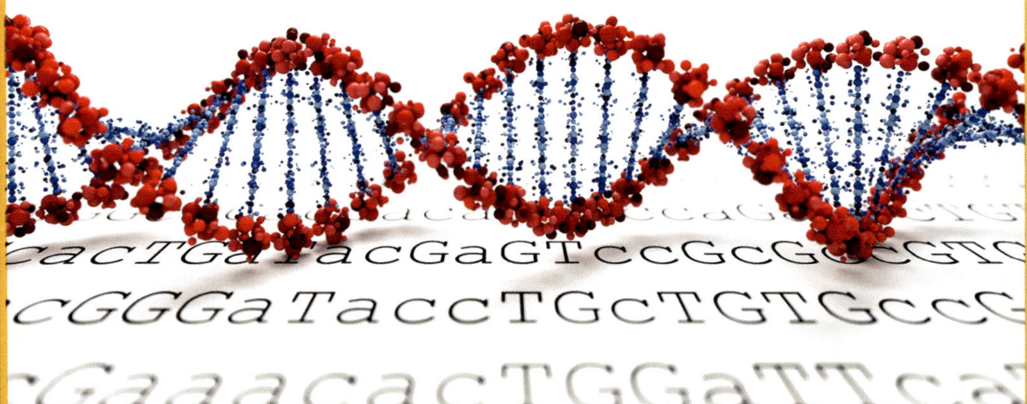

With these discoveries, geneticists no longer had to study genes indirectly by looking at phenotypes. In 1972, American Paul Berg made the first artificial recombinant DNA molecule. He spliced DNA from one species of bacteria into another species. Berg's discovery of recombinant DNA earned him the 1980 Nobel Prize in Chemistry. Scientists could now study actual genes, manipulate them, and even clone them. The era of genetic engineering had begun.

Geneticists realized the many beneficial implications of DNA research, including the promise of new and better foods and drugs. A gene from another organism can now be removed and spliced into a crop plant to make it more nutritious or disease-resistant. A cow can be engineered to produce more milk. Foreign genes can be inserted into bacteria so that they can produce substances such as human insulin and human growth hormone.

Another promise of genetics is the ability to treat genetic disorders. This requires analyzing genomes, which are the complete set of all genes in all chromosomes of an organism. Understanding the genome of any organism requires long, painstaking work. It can be divided into two parts: gene sequencing and gene mapping.

SEQUENCING

A base sequence is a complete list of bases found in all genes on all chromosomes of an organism, in order. The four bases are called nucleotides, and include the molecules cytosine, guanine, thymine, and adenine. These can be abbreviated A, T, C, and G. A base sequence is a long list of these four letters with no spaces. They're not neatly divided into recognizable units, such as words and sentences in language. So, just looking at the sequence won't tell you what it means.

Regardless of the creature, its base sequences are always huge. The bacterium *E. coli* contains just 4,600,000 base pairs. Assuming 3,000 characters per page, it would take 1,533 pages to print it all. Thomas Hunt Morgan's tiny fruit fly contains 180,000,000 bases, which would require 60,000 pages. Most people would think the human genome, with 3 billion base pairs, has the most bases and genes. However, a species of grasshopper has 180 billion base pairs. A single-celled amoeba has about 670 billion. The size and complexity of a creature doesn't always reflect the number of chromosomes, genes, and base pairs it has.

Many of the non-coding, or "junk," sequences in a genome are repeated over and over. Organisms with the largest genomes seem to have the most junk DNA. Until recently, people thought it had no function. But now they realize that some of it has a variety of cell functions, including regulating aspects of protein synthesis.

THE HUMAN GENOME PROJECT

Scientists began to record the base sequences of important research organisms, including *E. coli*, yeast, and the fruit fly. It soon became clear that many genes are nearly identical in all creatures. Closely related organisms have more DNA in common. However, half of fruit fly DNA is identical to human DNA. Humans and mice have 85 percent of their DNA in common, and humans and chimpanzees have 98 percent. Not surprisingly, some identical genes act the same. Genes for smell or taste act the same in nematode worms as they do in humans. Studying immunity genes in chickens helps us understand human immunity.

The Human Genome Project, first proposed in 1985, set out to sequence the entire human genome. Many people considered this project impossible. Some even compared it to landing a man on the moon. In 1987, however, the first automated DNA sequencing computer became commercially available. Scientists began the Human Genome Project in 1990. In 2000, a rough draft of the human genome was published. A much improved and corrected version followed in 2003. In the following years, scientists continued to fill in the gaps. Thousands of species, most of them bacteria, have been completely sequenced. Many are disease-causing bacteria and fungi. Others include honeybees, mice, rats, dogs, rice, poplar trees, cows, chimpanzees, mosquitoes, and puffer fish.

MAPPING

In addition to base sequences, gene maps are needed to understand genomes. A gene map is similar to a city map in that it locates every gene precisely by placing it at the correct locus on the correct chromosome. You might compare locating a specific gene to locating a person in a large city with just a picture but no name, address, or telephone number.

In 1911, Alfred Sturtevant, one of Morgan's students, created the first genetic map. It was of *Drosophila's* X chromosome. Since then, gene maps are still basically straight lines with "markers," or locations of known genes or DNA sequences, identified along their lengths. Some are physical maps, showing the exact number of base pairs between markers. Others are gene-linkage maps. They show gene locations in relation to each other, relative distances between genes, and likely linkage patterns. Sturtevant had only phenotypes to determine gene locations. Geneticists are now able to locate unique, nonrepeating DNA sequences, which are most likely genes.

GENE TRACKING

To track the exact gene causing a genetic defect, scientists follow two basic steps:

I. FIND THE PROTEIN CAUSING THE DEFECT.

A. Study disorder phenotypes using family histories and pedigrees.
B. Compare karyotypes to find abnormal chromosomes.
C. Study proteins in affected individuals to locate those with defects.
D. Find structure of affected proteins (determine their amino acid sequence).

II. USE THE PROTEIN TO LOCATE THE GENE.

A. Work backward from the protein's amino acid sequence to find its DNA sequence.
B. Search a "DNA library" (database of known DNA sequences) to find DNA segments used in the specific cells or tissues affected by the genetic defect. This is a complex, multistep process.

GENETICS IN THE FUTURE

Scientists have come a long way in understanding genetics since Gregor Mendel's time, but we still have a long way to go. Sequencing the human genome is just the start. Determining how to use the abundance of information contained in it will be the work of many generations. Each segment in the DNA "library" must be identified as either part of a gene coding for a protein or a segment of non-coding DNA. Each new gene locus that is pinpointed is a marker that makes it easier to find more new genes.

Geneticists studying long strands of junk DNA have discovered that some of these repeating segments can be used to identify individuals. This is the basis of DNA fingerprinting, which is now used—among other things—to determine family relationships and identify criminal suspects.

Continuing research into the human genome—and genomes of many other organisms—will likely transform our future. Among other things, geneticists will use this information to map locations of genes causing specific disorders and develop treatments. They are tracing the history and evolution of life on Earth. They hope to increase understanding of how the body works, how we are related to other organisms, and how we can use genetics to benefit our culture and environment.

If you are interested in a career in science, genetics might be a good field to pursue. The future of genetics is sure to be filled with exciting discoveries.

GLOSSARY

ALLELE An alternative form of a gene, such as those for blue or brown eye color or round or wrinkled seeds.

CHROMOSOME Structure in the cell nucleus composed of a single DNA molecule and containing many genes along its length.

DIHYBRID CROSS A genetic cross in which two genes are followed and the phenotypes counted.

DIPLOID Having two sets of chromosomes, one set from each parent; this occurs in all body cells.

DOMINANT Condition in which one allele masks the presence of another (recessive) allele of the same gene.

GAMETE Sex cell (egg, or ovum, in females and sperm in males).

GENE The unit or factor that determines a trait; located in a DNA molecule (chromosome) inside every cell nucleus.

GENOME The complete set of all genes in all chromosomes of an organism.

GENOTYPE The pair of alleles that combine in an organism to produce a phenotype; for example, BB or Bb for brown eye color and bb for blue eye color.

HAPLOID Having one set of chromosomes, which occurs in gametes as a result of meiosis (reduction division).

HETEROZYGOUS Having two different alleles in the genotype; for example, Yy.

HOMOLOGOUS CHROMOSOME Member of a pair of chromosomes in a diploid organism; homologous chromosome pairs are identical in size, shape, and sequence of genes.

HOMOZYGOUS Having two identical alleles in the genotype; for example, YY or yy.

HYBRID An offspring produced by crossing two genetically different parents of different types, which combines traits of both parents.

MEIOSIS The nuclear division process used to form gametes. This includes a reduction division, which halves the chromosome number and increases variation in offspring.

MITOSIS The nuclear division process used in the growth, maintenance, and repair of organisms; results in production of two identical daughter cells from a single parent cell.

MONOHYBRID CROSS A genetic cross in which only one trait (gene) is followed and the phenotypes counted.

MUTATION Alteration in the genetic makeup (DNA sequence) of an organism, which affects the genotype and often the phenotype.

NATURAL SELECTION The primary process by which evolution occurs; results when certain genetic types in a population are more successful than others in contributing offspring to the next generation.

PHENOTYPE The visible traits of an organism, such as eye color, seed color, or seed type.

PROBABILITY The chance, or likelihood, that a given event will occur; can be expressed as a fraction, decimal, or percentage (for example, ¼, 0.25, or 25 percent).

PUNNETT SQUARE Diagram illustrating the results of a cross by showing all possible combinations of genotype.

RECESSIVE Condition in which the expression of one allele is masked by the presence of another (dominant) allele of the same gene.

SELECTIVE BREEDING Artificially influencing the inheritance of traits by selecting the parents of each new generation.

TESTCROSS A cross between an individual displaying the dominant phenotype and a homozygous recessive individual, which is used to determine an individual's genotype.

FOR MORE INFORMATION

CANADIAN ASSOCIATION OF GENETIC COUNSELLORS
P.O. Box 52083
Oakville, ON L6J 7N5
Canada
(905) 847-1363
Website: www.cagc-accg.ca
This organization for professional genetic counselors and students seeks to increase understanding and awareness of the field. The website includes information on the purposes of genetic counseling and links to related sites.

CHARLES DARWIN FOUNDATION
Charles Darwin Research Station
Av. Charles Darwin s/n
Puerto Ayora, Galápagos, Ecuador
(593) 5 2526-146/2527-013 Ext 101
Website: www.darwinfoundation.org/en/
The Charles Darwin Foundation is a nonprofit organization dedicated to responsible scientific research on the Galápagos Islands since 1959.

GENETICS SOCIETY OF AMERICA (GSA)
6120 Executive Boulevard
Suite 550
Rockville, MD 20852
(240) 880-2000
Website: www.genetics-gsa.org
This organization brings together geneticists, promotes research in genetics, helps train new geneticists, and educates the public and government about advances in genetics and their potential consequences.

GENOME CANADA
150 Metcalfe Street, Suite 2100

Ottawa, ON K2P 1P1
Canada
(613) 751-4460
Website: genomecanada.ca/en
Genome Canada is a nonprofit organization dedicated to advancing genome research for the benefit of all Canadians.

GENOMIC SCIENCE PROGRAM

U.S. Department of Energy
1000 Independence Avenue SW
Washington, DC 20585
(202) 586-5000
Website: genomics.energy.gov
The U.S. Department of Energy is one of the founders of the Human Genome Project and runs many genetics projects. The site is filled with images and educational materials.

NATIONAL HUMAN GENOME RESEARCH INSTITUTE (NHGRI)

Communications and Public Liaison Branch
National Institutes of Health
Building 31, Room 4B09
31 Center Drive, MSC 2152
9000 Rockville Pike
Bethesda, MD 20892-2152
(301) 402-0911
Website: www.genome.gov
The NHGRI led the project to sequence the human genome and is now leading the research to understand its structure, function, and role in health and disease.

POSITIVE EXPOSURE 109 GALLERY AND PERFORMANCE SPACE

14 East 109th Street
New York, NY 10029
(212) 420-1931
Website: www.positiveexposure.org
This arts organization was founded by a former fashion photographer and an M.D./Ph.D. to work with individuals and families living with genetic differences. Its gives workshops, educational and human-rights programs, and multimedia exhibitions to educate physicians, health care workers, schools, and the public about genetic differences.

FOR FURTHER READING

Darwin, Charles. *The Origin of Species by Means of Natural Selection.* New York, NY: SNOVA, 2019.

Folkersen, Lasse. *Understand Your DNA: A Guide.* Hackensack, NJ: World Scientific, 2019.

Fuller, Randall. T*he Book That Changed America: How Darwin's Theory of Evolution Ignited a Nation.* New York, NY: Viking, 2017.

Irving, Melita, Dr. T*he Human DNA Manual: Understanding Your Genetic Code.* Newbury Park, CA: Haynes Publishing, 2019.

Lewis, Ricki. Human *Genetics: The Basics.* New York, NY: Routledge, 2017.

Mendel, Gregor. *Experiments in Plant Hybridisation* (Illustrated). Createspace Independent Publishing Platform, 2017.

Neal, Dick. *Introduction to Population Biology.* New York, NY: Cambridge University Press, 2019.

Rauf, Don. *DNA, RNA, and the Inheritance of Traits.* New York, NY: Enslow Publishing, 2018.

Roberts, Alice, et al. *Evolution: The Human Story.* New York, NY: DK Publishing, 2018.

Robinson, Tara Rodden. *Genetics for Dummies* (third edition). Hoboken, NJ: John Wiley & Sons, 2020.

Sneddon, Bradley. *Charles Darwin*. Broomall, PA: Mason Crest, 2019.

Strager, Hanne. *A Modest Genius: The Story of Darwin's Life and How His Ideas Changed Everything*. Createspace Independent Publishing Platform, 2016.

Wilmer, George. *Gregor Mendel*. Broomall, PA: Mason Crest Publishers, 2018.

INDEX

M

N

O

P

R

S

T

PHOTO CREDITS

Cover, pp.1–80 Who is Danny/Shutterstock.com; Cover, pogonici/Shutterstock.com; p. 4 Tartila/Shutterstock.com; p. 5 Monkey Business Images/Shutterstock.com; p. 7 Marzolino/Shutterstock.com; p. 9 Heritage Images/Contributor/Hulton Archive/Getty Images; p. 11 SanderMeertinsPhotography/Shutterstock.com; p. 13 Aldona Griskeviciene/Shutterstock.com; p. 14 Universal History Archive/Contributor/Universal Images Group Editorial/Getty Images; p. 17 ttsz/iStock/Getty Images; p. 19 N.Vinoth Narasingam/Shutterstock.com; p. 22 https://en.wikipedia.org/wiki/File:Reginald_Crundall_Punnett,_1875%E2%80%931967.jpg; p. 25 Natali _ Mis/Shutterstock.com; p 29 VectorMine/Shutterstock.com; p. 31 vdvornyk/ iStock Editorial/Getty Images; p. 33 Victoria Kurylo/Shutterstock.com; p. 34 https://commons.wikimedia.org/wiki/File:Co-dominance_Rhododendron.jpg; p.35 NOTE OMG/Shutterstock.com; p. 37 Roger Givens/Shutterstock.com; p. 39 Africa Studio/Shuttetstock.com; p. 40 gritsalak karalak/Shuttestock.com; p. 41 Iaremenko Sergii/Shutterstock.com; p. 42 Fancy Tapis/Shutterstock.com; p. 44 Rattiya Thongdumhyu/Shutterstock.com; p. 45 Mark_Kostich/Shutterstock.com; p. 47 nobeastsofierce/Shutterstock.com; p. 49 Rudy Umans/Shutterstock.com; pp. 51, 53 wanpiya Chaikitwattana/Shutterstock.com; p. 55 BSIP/Contributor/ Universal Images Group/Getty Images; p. 57 somersault1824/Shutterstock.com; p. 59 Alexlky/Shutterstock.com; p. 63 https://commons.wikimedia.org/wiki/File:Morgan_crossover_1.jpg; p. 64 Leigh Prather/Shutterstock.com; p. 69 Maksim Shmeljov/Shutterstock.com.